Student Resources

M000105166

© Houghton Mifflin Harcourt Publishing Company • Cover Image Credits: (Hares) ©Radius Images/Corbis; (Garden, New York) ©Rick Lew/The Image Bank/Getty Images; (sky) ©PhotoDisc/Getty Images

Harcourt

INCLUDES
- Program Authors
- Table of Contents
- Picture Glossary
- Common Core State Standards Correlation
- Index

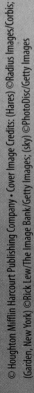

Made in the United States
Text printed on 100%
recycled paper

Houghton Mifflin Harcourt

Copyright © by Houghton Mifflin Harcourt Publishing Company

All rights reserved. No part of this work may be reproduced or transmitted in any form or by any means, electronic or mechanical, including photocopying or recording, or by any information storage or retrieval system, without the prior written permission of the copyright owner unless such copying is expressly permitted by federal copyright law.

Permission is hereby granted to individuals using the corresponding student's textbook or kit as the major vehicle for regular classroom instruction to photocopy entire pages from this publication in classroom quantities for instructional use and not for resale. Requests for information on other matters regarding duplication of this work should be addressed to Houghton Mifflin Harcourt Publishing Company, Attn: Contracts, Copyrights, and Licensing, 9400 Southpark Center Loop, Orlando, Florida 32819-8647.

Common Core State Standards © Copyright 2010. National Governors Association Center for Best Practices and Council of Chief State School Officers. All rights reserved.

This product is not sponsored or endorsed by the Common Core State Standards Initiative of the National Governors Association Center for Best Practices and the Council of Chief State School Officers.

Printed in the U.S.A.

ISBN 978-0-544-34343-6

18 0928 19

4500788027   B C D E F G

If you have received these materials as examination copies free of charge, Houghton Mifflin Harcourt Publishing Company retains title to the materials and they may not be resold. Resale of examination copies is strictly prohibited.

Possession of this publication in print format does not entitle users to convert this publication, or any portion of it, into electronic format.

Dear Students and Families,

Welcome to **Go Math!**, Grade K! In this exciting mathematics program, there are hands-on activities to do and real-world problems to solve. Best of all, you will write your ideas and answers right in your book. In **Go Math!**, writing and drawing on the pages helps you think deeply about what you are learning, and you will really understand math!

By the way, all of the pages in your **Go Math!** book are made using recycled paper. We wanted you to know that you can Go Green with **Go Math!**

Sincerely,

The Authors

Made in the United States
Text printed on 100% recycled paper

© Houghton Mifflin Harcourt Publishing Company • Image Credits: (bg) ©Sankar Salvady/Flickr/Getty Images; (t) ©Blaine Harrington III/Alamy Images; (c) ©Don Johnston/All Canada Photos/Getty Images; (b) ©Erich Kuchling/Westend61/Corbis

GO MATH!

Authors

Juli K. Dixon, Ph.D.
Professor, Mathematics Education
University of Central Florida
Orlando, Florida

Edward B. Burger, Ph.D.
President, Southwestern University
Georgetown, Texas

Steven J. Leinwand
Principal Research Analyst
American Institutes for
  Research (AIR)
Washington, D.C.

Contributor

Rena Petrello
Professor, Mathematics
Moorpark College
Moorpark, CA

Matthew R. Larson, Ph.D.
K-12 Curriculum Specialist for
  Mathematics
Lincoln Public Schools
Lincoln, Nebraska

Martha E. Sandoval-Martinez
Math Instructor
El Camino College
Torrance, California

English Language Learners Consultant

Elizabeth Jiménez
CEO, GEMAS Consulting
Professional Expert on English
  Learner Education
Bilingual Education and
  Dual Language
Pomona, California

© Houghton Mifflin Harcourt Publishing Company • Image Credits: (t) ©Richard Wear/Design Pics/Corbis; (bg) ©Russ Bishop/Alamy Images

Table of Contents

Number and Operations

Critical Area Representing, relating, and operating on whole numbers, initially with sets of objects

1 Represent, Count, and Write Numbers 0 to 5   9

Domains Counting and Cardinality
        Operations and Algebraic Thinking
COMMON CORE STATE STANDARDS
K.CC.A.3, K.CC.B.4a, K.CC.B.4b, K.CC.B.4c, K.OA.A.3

2 Compare Numbers to 5   77

Domain Counting and Cardinality
COMMON CORE STATE STANDARDS
K.CC.C.6

© Houghton Mifflin Harcourt Publishing Company

Critical Area

GO DIGITAL

Go online! Your math lessons are interactive. Use *iTools*, Animated Math Models, the Multimedia *eGlossary*, and more.

Chapter 1 Overview

In this chapter, you will explore and discover answers to the following **Essential Questions**:

- How can you show, count, and write numbers?
- How can you show numbers 0 to 5?
- How can you count numbers 0 to 5?
- How can you write numbers 0 to 5?

Chapter 2 Overview

In this chapter, you will explore and discover answers to the following **Essential Questions**:

- How can building and comparing sets help you compare numbers?
- How does matching help you compare sets?
- How does counting help you compare sets?
- How do you know if the number of counters in one set is the same as, greater than, or less than the number of counters in another set?

v

Chapter 3 Overview

In this chapter, you will explore and discover answers to the following **Essential Questions**:

• How can you show, count, and write numbers 6 to 9?

• How can you show numbers 6 to 9?

• How can you count numbers 6 to 9?

• How can you write numbers 6 to 9?

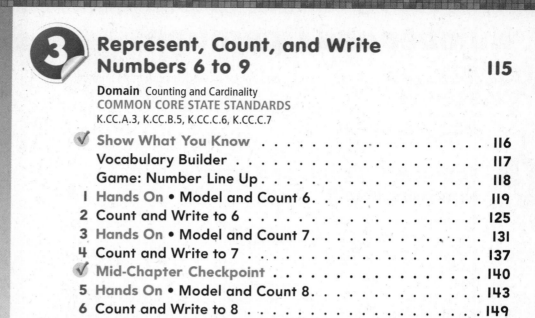

3 Represent, Count, and Write Numbers 6 to 9   115

Domain Counting and Cardinality
COMMON CORE STATE STANDARDS
K.CC.A.3, K.CC.B.5, K.CC.C.6, K.CC.C.7

Practice and Homework

Lesson Check and Spiral Review in every lesson

vi

© Houghton Mifflin Harcourt Publishing Company

4 Represent and Compare Numbers to 10 — 177

Domains Counting and Cardinality
Operations and Algebraic Thinking
COMMON CORE STATE STANDARDS
K.CC.A.2, K.CC.A.3, K.CC.B.5, K.CC.C.6, K.CC.C.7, K.OA.A.3, K.OA.A.4

Chapter 4 Overview

In this chapter, you will explore and discover answers to the following **Essential Questions:**
• How can you show and compare numbers to 10?
• How can you count forward to 10?
• How can you show numbers from 1 to 10?
• How can using models help you compare two numbers?

5 Addition — 227

Domain Operations and Algebraic Thinking
COMMON CORE STATE STANDARDS
K.OA.A.1, K.OA.A.2, K.OA.A.3, K.OA.A.4, K.OA.A.5

Chapter 5 Overview

In this chapter, you will explore and discover answers to the following **Essential Questions:**
• How can you show addition?
• How can using objects or pictures help you show addition?
• How can you use numbers and symbols to show addition?

Personal Math Trainer
Online Assessment and Intervention

© Houghton Mifflin Harcourt Publishing Company

Chapter 6 Overview

In this chapter, you will explore and discover answers to the following **Essential Questions**:

• How can you show subtraction?

• How can you use numbers and symbols to show a subtraction sentence?

• How can using objects and drawings help you solve word problems?

• How can acting it out help you solve subtraction word problems?

• How can using addition help you solve subtraction word problems?

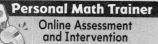

Personal Math Trainer
Online Assessment and Intervention

6 Subtraction 307

Domain Operations and Algebraic Thinking
COMMON CORE STATE STANDARDS
K.OA.A.1, K.OA.A.2, K.OA.A.5

© Houghton Mifflin Harcourt Publishing Company

Chapter 7 Overview

In this chapter, you will explore and discover answers to the following **Essential Questions**:

• How can you show, count, and write numbers 11 to 19?
• How can you show numbers 11 to 19?
• How can you read and write numbers 11 to 19?
• How can you show the teen numbers as 10 and some more?

Chapter 8 Overview

In this chapter, you will explore and discover answers to the following **Essential Questions**:

• How can you show, count, and write numbers to 20 and beyond?
• How can you show and write numbers to 20?
• How can you count numbers to 50 by ones?
• How can you count numbers to 100 by tens?

© Houghton Mifflin Harcourt Publishing Company

Geometry and Positions

Common Core **Critical Area** Describing shapes and space

9 **Identify and Describe Two-Dimensional Shapes** **489**

Domain Geometry
COMMON CORE STATE STANDARDS
K.G.A.2, K.G.B.4, K.G.B.6

GO DIGITAL

Go online! Your math lessons are interactive. Use *iTools*, Animated Math Models, the Multimedia *eGlossary*, and more.

Chapter 9 Overview

In this chapter, you will explore and discover answers to the following **Essential Questions**:

• How can you identify, name, and describe two-dimensional shapes?

• How can knowing the parts of two-dimensional shapes help you join shapes?

• How can knowing the number of sides and vertices of two-dimensional shapes help you identify shapes?

Personal Math Trainer
Online Assessment and Intervention

© Houghton Mifflin Harcourt Publishing Company

10 Identify and Describe Three-Dimensional Shapes   569

Domain Geometry

COMMON CORE STATE STANDARDS
K.G.A.1, K.G.A.2, K.G.A.3, K.G.B.4, K.G.B.5

Chapter 10 Overview

In this chapter, you will explore and discover answers to the following **Essential Questions**:

• How can identifying and describing shapes help you sort them?
• How can you describe three-dimensional shapes?
• How can you sort three-dimensional shapes?

Practice and Homework

Lesson Check and Spiral Review in every lesson

© Houghton Mifflin Harcourt Publishing Company

GO DIGITAL

Go online! Your math lessons are interactive. Use *i*Tools, Animated Math Models, the Multimedia *e*Glossary, and more.

Chapter 11 Overview

In this chapter, you will explore and discover answers to the following **Essential Questions**:

- How can comparing objects help you measure them?
- How can you compare the length of objects?
- How can you compare the height of objects?
- How can you compare the weight of objects?

Chapter 12 Overview

In this chapter, you will explore and discover answers to the following **Essential Questions**:

- How does sorting help you display information?
- How can you sort and classify objects by color?
- How can you sort and classify objects by shape?
- How can you sort and classify objects by size?
- How do you display information on a graph?

Measurement and Data

Common Core **Critical Area** Representing, relating, and operating on whole numbers, initially with sets of objects

11 Measurement 645

Domain Measurement and Data
COMMON CORE STATE STANDARDS
K.MD.A.1, K.MD.A.2

12 Classify and Sort Data 683

Domain Measurement and Data
COMMON CORE STATE STANDARDS
K.MD.B.3

© Houghton Mifflin Harcourt Publishing Company

Picture Glossary

above [arriba, encima]

The kite is **above** the rabbit.

add [sumar]

$$3 + 2 = 5$$

alike [igual]

and [y]

and

$$2 + 2$$

behind [detrás]

The box is **behind** the girl.

below [debajo]

The rabbit is **below** the kite.

beside [al lado]

The tree is **beside** the bush.

© Houghton Mifflin Harcourt Publishing Company

big [grande]

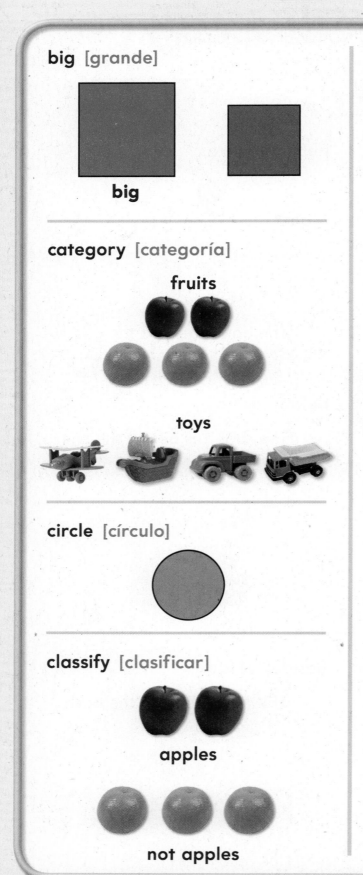

big

category [categoría]

fruits

toys

circle [círculo]

classify [clasificar]

apples

not apples

color [color]

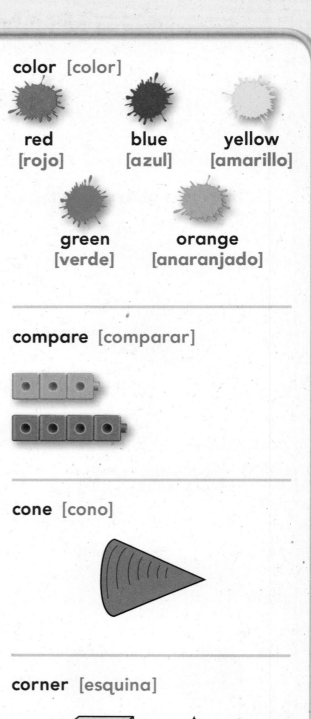

red
[rojo]

blue
[azul]

yellow
[amarillo]

green
[verde]

orange
[anaranjado]

compare [comparar]

cone [cono]

corner [esquina]

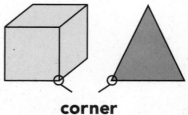

corner

© Houghton Mifflin Harcourt Publishing Company • Image Credits: (apples) ©Artville/Getty Images; (boat) ©D. Hurst/Alamy; (blue truck) ©C Squared Studios/PhotoDisc/Getty Images

cube [cubo]

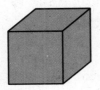

curve [curva]

curved surface
[superficie curva]

Some solids have
a **curved surface**.

cylinder [cilindro]

different [diferente]

eight [ocho]

eighteen [dieciocho]

eleven [once]

fewer [menos]

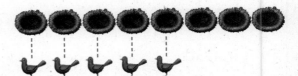

3 **fewer** birds

© Houghton Mifflin Harcourt Publishing Company

fifteen [quince]

fifty [cincuenta]

1	2	3	4	5	6	7	8	9	10
11	12	13	14	15	16	17	18	19	20
21	22	23	24	25	26	27	28	29	30
31	32	33	34	35	36	37	38	39	40
41	42	43	44	45	46	47	48	49	50

five [cinco]

flat [plano]

A circle is a **flat** shape.

flat surface [superficie plana]

Some solids have a flat **surface**.

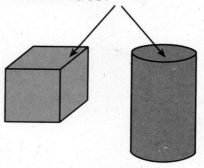

four [cuatro]

fourteen [catorce]

graph [gráfica]

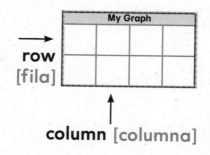

row [fila]

column [columna]

© Houghton Mifflin Harcourt Publishing Company

greater [mayor]

9 is greater than 6

6

9

heavier [más pesado]

heavier

hexagon [hexágono]

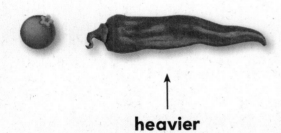

in front of [delante de]

The box is **in front of** the girl.

is equal to [es igual a]

$3 + 2 = 5$

$3 + 2$ **is equal to** 5

larger [más grande]

2   3

A quantity of 3 is **larger** than a quantity of 2.

less [menor/menos]

9 is **less** than 11

9

11

lighter [más liviano]

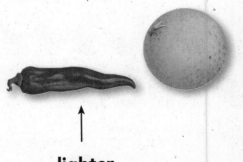

lighter

© Houghton Mifflin Harcourt Publishing Company

longer [más largo]

 **longer**

match [emparejar]

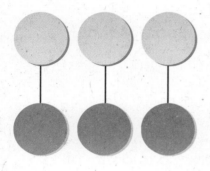

minus – [menos]

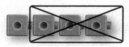

4 – 3 = 1

4 **minus** 3 is equal to 1

more [más]

2 **more** leaves

next to [al lado de]

The bush is **next to** the tree.

nine [nueve]

nineteen [diecinueve]

one [uno]

© Houghton Mifflin Harcourt Publishing Company

one hundred [cien]

1	2	3	4	5	6	7	8	9	10
11	12	13	14	15	16	17	18	19	20
21	22	23	24	25	26	27	28	29	30
31	32	33	34	35	36	37	38	39	40
41	42	43	44	45	46	47	48	49	50
51	52	53	54	55	56	57	58	59	60
61	62	63	64	65	66	67	68	69	70
71	72	73	74	75	76	77	78	79	80
81	82	83	84	85	86	87	88	89	90
91	92	93	94	95	96	97	98	99	100

ones [unidades]

3 **ones**

pairs [pares]

3

3	0
2	1
1	2
0	3

number **pairs** for 3

plus + [más]

2 **plus** 1 is equal to 3

$2 + 1 = 3$

rectangle [rectángulo]

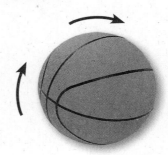

roll [rodar]

same height
[de la misma altura]

© Houghton Mifflin Harcourt Publishing Company

same length [del mismo largo]

seventeen [diecisiete]

same number
[el mismo número]

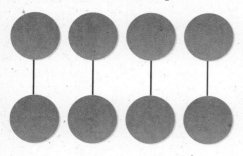

shape [forma]

same weight [del mismo peso]

shorter [más corto]

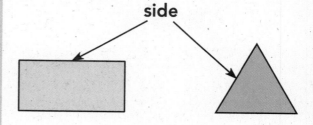

shorter

seven [siete]

side [lado]

side

© Houghton Mifflin Harcourt Publishing Company

sides of equal length [lados del mismo largo]

six [seis]

sixteen [dieciséis]

size [tamaño]

big   small

slide [deslizar]

small [pequeño]

small

solid [sólido]

solid

A cylinder is a **solid** shape.

© Houghton Mifflin Harcourt Publishing Company • Image Credits: (tr) ©Eyewire/Getty Images

sphere [esfera]

square [cuadrado]

stack [apilar]

subtract [restar]

Subtract to find out how many are left.

taller [más alto]

taller

ten [diez]

tens [decenas]

1	2	3	4	5	6	7	8	9	10
11	12	13	14	15	16	17	18	19	20
21	22	23	24	25	26	27	28	29	30
31	32	33	34	35	36	37	38	39	40
41	42	43	44	45	46	47	48	49	50
51	52	53	54	55	56	57	58	59	60
61	62	63	64	65	66	67	68	69	70
71	72	73	74	75	76	77	78	79	80
81	82	83	84	85	86	87	88	89	90
91	92	93	94	95	96	97	98	99	100

tens

© Houghton Mifflin Harcourt Publishing Company

thirteen [trece]

three [tres]

three-dimensional shapes
[figuras tridimensionales]

triangle [triángulo]

twelve [doce]

twenty [veinte]

two [dos]

two-dimensional shapes
[figuras bidimensionales]

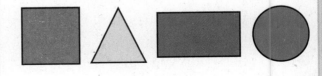

© Houghton Mifflin Harcourt Publishing Company

vertex [vértice]

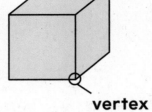

vertex

---

vertices [vértices]

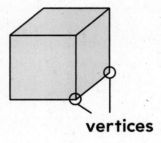

vertices

---

zero, none [cero, ninguno]

zero fish

© Houghton Mifflin Harcourt Publishing Company • Image Credits: (tr) ©C Squared Studios/PhotoDisc/Getty Images

Correlations

COMMON CORE STATE STANDARDS

Standards You Will Learn

Mathematical Practices		Some examples are:
MP1	Make sense of problems and persevere in solving them.	Lessons 1.3, 1.5, 1.9, 3.9, 5.1, 5.3, 5.4, 5.6, 5.7, 6.1, 6.3, 6.4, 6.5, 6.6, 7.6, 11.3, 11.5
MP2	Reason abstractly and quantitatively.	Lessons 1.1, 1.2, 1.3, 1.4, 1.5, 1.6, 1.8, 1.9, 1.10, 2.2, 2.3, 2.5, 3.2, 3.4, 3.6, 3.8, 4.2, 4.4, 5.1, 5.2, 5.3, 5.4, 5.5, 5.6, 5.7, 5.8, 5.9, 5.10, 5.11, 5.12, 6.1, 6.2, 6.3, 6.4, 6.5, 6.6, 6.7, 7.1, 7.2, 7.3, 7.4, 7.5, 7.6, 7.7, 7.8, 7.9, 7.10, 8.1, 8.2, 8.3, 9.4, 9.6, 9.8, 9.10, 10.3, 10.4, 10.5, 12.1, 12.2, 12.3, 12.4, 12.5
MP3	Construct viable arguments and critique the reasoning of others.	Lessons 2.1, 2.2, 2.3, 2.4, 2.5, 3.9, 7.1, 7.3, 7.7, 7.9, 10.7, 10.9, 10.10, 11.1, 11.2, 11.3, 11.4, 11.5
MP4	Model with mathematics.	Lessons 1.7, 1.9, 2.4, 3.1, 3.9, 4.1, 4.3, 4.5, 5.2, 5.3, 5.4, 6.2, 6.3, 6.4, 7.6, 8.4, 10.6, 10.8, 10.9, 10.10
MP5	Use appropriate tools strategically.	Lessons 1.8, 2.1, 2.2, 2.3, 2.4, 3.1, 3.3, 3.5, 3.7, 4.1, 4.5, 5.2, 6.2, 6.7, 7.5, 8.1, 8.4, 9.1, 9.2, 9.3, 9.5, 9.7, 9.9, 9.11, 9.12, 10.1, 10.2, 10.3, 10.4, 10.5, 10.6, 11.1, 11.2, 11.4, 12.1, 12.2, 12.3
MP6	Attend to precision.	Lessons 2.5, 4.6, 4.7, 8.1, 8.7, 9.1, 9.3, 9.5, 9.7, 9.9, 10.1, 10.2, 10.3, 10.4, 10.5, 10.9, 10.10, 11.1, 11.2, 11.3, 11.4, 11.5, 12.1, 12.2, 12.3, 12.4, 12.5
MP7	Look for and make use of structure.	Lessons 1.7, 1.8, 3.1, 3.3, 3.5, 3.7, 4.3, 5.5, 5.8, 5.9, 5.10, 5.11, 5.12, 7.1, 7.2, 7.3, 7.4, 7.5, 7.7, 7.8, 7.9, 7.10, 8.5, 8.6, 8.7, 8.8, 9.1, 9.2, 9.3, 9.4, 9.5, 9.6, 9.7, 9.8, 9.9, 9.10, 9.11, 9.12, 10.1, 10.2, 10.6

© Houghton Mifflin-Harcourt Publishing Company

Standards You Will Learn

Mathematical Practices		Some examples are:
MP8	Look for and express regularity in repeated reasoning.	Lessons 3.3, 3.5, 3.7, 4.5, 4.6, 4.7, 5.5, 6.7, 7.2, 7.4, 7.8, 7.10, 8.5, 8.6, 8.7, 8.8, 9.4, 9.6, 9.8, 9.10, 9.11, 9.12, 10.7, 12.4, 12.5
Domain: Counting and Cardinality		**Student Edition Lessons**
Know number names and the count sequence.		
K.CC.A.1	Count to 100 by ones and by tens.	Lessons 8.5, 8.6, 8.7, 8.8
K.CC.A.2	Count forward beginning from a given number within the known sequence (instead of having to begin at 1).	Lessons 4.4, 8.3, 8.5
K.CC.A.3	Write numbers from 0 to 20. Represent a number of objects with a written numeral 0–20 (with 0 representing a count of no objects).	Lessons 1.2, 1.4, 1.6, 1.9, 1.10, 3.2, 3.4, 3.6, 3.8, 4.2, 8.2
Count to tell the number of objects.		
K.CC.B.4a	Understand the relationship between numbers and quantities; connect counting to cardinality. a. When counting objects, say the number names in the standard order, pairing each object with one and only one number name and each number name with one and only one object.	Lessons 1.1, 1.3, 1.5
K.CC.B.4b	Understand the relationship between numbers and quantities; connect counting to cardinality. b. Understand that the last number name said tells the number of objects counted. The number of objects is the same regardless of their arrangement or the order in which they were counted.	Lesson 1.7
K.CC.B.4c	Understand the relationship between numbers and quantities; connect counting to cardinality. c. Understand that each successive number name refers to a quantity that is one larger.	Lesson 1.8

© Houghton Mifflin Harcourt Publishing Company

Standards You Will Learn

Domain: Counting and Cardinality

Count to tell the number of objects.

K.CC.B.5	Count to answer "how many?" questions about as many as 20 things arranged in a line, a rectangular array, or a circle, or as many as 10 things in a scattered configuration; given a number from 1–20, count out that many objects.	Lessons 3.1, 3.3, 3.5, 3.7, 4.1, 8.1

Compare numbers.

K.CC.C.6	Identify whether the number of objects in one group is greater than, less than, or equal to the number of objects in another group, e.g., by using matching and counting strategies.	Lessons 2.1, 2.2, 2.3, 2.4, 2.5, 3.9, 4.5, 4.6, 8.4
K.CC.C.7	Compare two numbers between 1 and 10 presented as written numerals.	Lessons 3.9, 4.7, 8.6

Domain: Operations and Algebraic Thinking

Understand addition as putting together and adding to, and understand subtraction as taking apart and taking from.

K.OA.A.1	Represent addition and subtraction with objects, fingers, mental images, drawings, sounds (e.g., claps), acting out situations, verbal explanations, expressions, or equations.	Lessons 5.1, 5.2, 5.3, 6.1, 6.2, 6.3
K.OA.A.2	Solve addition and subtraction word problems, and add and subtract within 10, e.g., by using objects or drawings to represent the problem.	Lessons 5.7, 6.6, 6.7
K.OA.A.3	Decompose numbers less than or equal to 10 into pairs in more than one way, e.g., by using objects or drawings, and record each decomposition by a drawing or equation (e.g., 5 = 2 + 3 and 5 = 4 + 1).	Lessons 1.7, 4.1, 5.8, 5.9, 5.10, 5.11, 5.12

© Houghton Mifflin Harcourt Publishing Company

Standards You Will Learn

Domain: Operations and Algebraic Thinking		
Understand addition as putting together and adding to, and understand subtraction as taking apart and taking from.		
K.OA.A.4	For any number from 1 to 9, find the number that makes 10 when added to the given number, e.g., by using objects or drawings, and record the answer with a drawing or equation.	Lessons 4.3, 5.5
K.OA.A.5	Fluently add and subtract within 5.	Lessons 5.4, 5.6, 6.4, 6.5
Domain: Number and Operations in Base Ten		
Work with numbers 11–19 to gain foundations for place value.		
K.NBT.A.1	Compose and decompose numbers from 11 to 19 into ten ones and some further ones, e.g., by using objects or drawings, and record each composition or decomposition by a drawing or equation (e.g., 18 = 10 + 8); understand that these numbers are composed of ten ones and one, two, three, four, five, six, seven, eight, or nine ones.	Lessons 7.1, 7.2, 7.3, 7.4, 7.5, 7.7, 7.8, 7.9, 7.10
Domain: Measurement and Data		
Describe and compare measurable attributes.		
K.MD.A.1	Describe measurable attributes of objects, such as length or weight. Describe several measurable attributes of a single object.	Lesson 11.5
K.MD.A.2	Directly compare two objects with a measurable attribute in common, to see which object has "more of"/ "less of" the attribute, and describe the difference.	Lessons 11.1, 11.2, 11.3, 11.4

Common Core State Standards © Copyright 2010. National Governors Association Center for Best Practices and Council of Chief State School Officers. All rights reserved. This product is not sponsored or endorsed by the Common Core State Standards Initiative of the National Governors Association Center for Best Practices and the Council of Chief State School Officers.

© Houghton Mifflin Harcourt Publishing Company

Standards You Will Learn

Domain: Measurement and Data		
Classify objects and count the number of objects in each category.		
K.MD.B.3	Classify objects into given categories; count the numbers of objects in each category and sort the categories by count.	Lessons 12.1, 12.2, 12.3, 12.4, 12.5
Domain: Geometry		
Identify and describe shapes (squares, circles, triangles, rectangles, hexagons, cubes, cones, cylinders, and spheres).		
K.G.A.1	Describe objects in the environment using names of shapes, and describe the relative positions of these objects using terms such as *above, below, beside, in front of, behind,* and *next to.*	Lessons 10.8, 10.9, 10.10
K.G.A.2	Correctly name shapes regardless of their orientations or overall size.	Lessons 9.1, 9.3, 9.5, 9.7, 9.9, 10.2, 10.3, 10.4, 10.5
K.G.A.3	Identify shapes as two-dimensional (lying in a plane, "flat") or three-dimensional ("solid").	Lesson 10.6
Analyze, compare, create, and compose shapes.		
K.G.B.4	Analyze and compare two- and three-dimensional shapes, in different sizes and orientations, using informal language to describe their similarities, differences, parts (e.g., number of sides and vertices / "corners") and other attributes (e.g., having sides of equal length).	Lessons 9.2, 9.4, 9.6, 9.8, 9.10, 9.11, 10.1
K.G.B.5	Model shapes in the world by building shapes from components (e.g., sticks and clay balls) and drawing shapes.	Lesson 10.7
K.G.B.6	Compose simple shapes to form larger shapes.	Lesson 9.12

© Houghton Mifflin Harcourt Publishing Company

Index

© Houghton Mifflin Harcourt Publishing Company

Circle
   curve, 499–502
   describe, 499–502
   identify and name, 493–496
   sort, 493–496

Classify
   and count by color, 687–690
   and count by shape, 693–696
   and count by size, 699–701

Color
   sort by, 687–690

Common Core State Standards, H13–H17

Compare
   by counting
      sets to 5, 105–108
      sets to 10, 211–214
   heights, 655–663, 673–676
   lengths, 649–652, 673–676
      by matching
      sets to 5, 99–102
      sets to 10, 205–208
   numbers/sets
      greater, 87–90
      less, 93–95
      same, 81–84
      to five, 99–102, 105–108
      to ten, 205–208, 211–214
      to twenty, 447–449
   two-dimensional shapes, 553–556
   two numbers, 217–220
   weights, 667–670, 673–676

Cone
   curved surface, 597–599
   flat surface, 598
   identify, name, and describe, 597–599
   sort, 597

Corners. *See* Vertices

Correlations
   Common Core State Standards,
      H13–H19

Count
   compare by, 105–108, 211–214
   forward
      to fifty, 453–456
      to one hundred, 459–462, 465–468,
         471–474
      to twenty, 429–432, 435–438,
         441–444

   model and, 13–16, 25–28, 37–40,
      67–70, 119–122, 131–134, 143–146,
      155–158, 181–184, 361–364,
      373–376, 385–388, 397–400,
      409–412, 429–432
   and write, 19–22, 31–33, 43–46,
      67–70, 125–128, 137–139,
      149–152, 161–164, 187–190,
      367–369, 379–382, 385–388,
      403–406, 415–418, 435–438
   numbers. *See* Numbers
   by ones, 453–456, 459–462
   by tens, 465–468, 471–474

Cube
   flat surfaces, 586
   identify, name, and describe, 585–588
   sort, 585–588

Curious George®, Curious About Math,
   9, 77, 115, 177, 227, 307, 357, 425,
   489, 569, 645, 683

Curve
   of circle, 499–502
   sort by, 553–556

Curved surface, 580, 592, 598

Cylinder
   curved surface, 592
   flat surfaces, 592
   identify, name, and describe, 591–594
   sort, 591–594

D

Data
   classify and count, 687–690, 693–696,
      699–701
   graphs, concrete, 705–708, 711–714

Different, 553–556

E

Eight
   count, 143–146, 149–152
   model, 143–146
   write, 149–152

© Houghton Mifflin Harcourt Publishing Company

© Houghton Mifflin Harcourt Publishing Company

© Houghton Mifflin Harcourt Publishing Company

3. Construct viable arguments and critique the reasoning of others. In many lessons. Some examples are: 81, 87, 93, 99, 105, 167, 361, 373, 397, 409, 447, 609, 621, 627, 649, 655, 661, 667, 673

4. Model with mathematics. In many lessons. Some examples are: 49, 61, 99, 119, 167, 181, 193, 205, 237, 243, 249, 317, 323, 329, 391, 447, 541, 603, 615, 621, 627

5. Use appropriate tools strategically. In many lessons. Some examples are: 55, 81, 87, 93, 99, 119, 131, 143, 155, 181, 205, 237, 317, 347, 385, 429, 447, 493, 499, 505, 517, 529, 541, 553, 559, 573, 579, 585, 591, 597

6. Attend to precision. In many lessons. Some examples are: 105, 211, 217, 429, 465, 493, 499, 505, 517, 529, 573, 579, 585, 591, 597, 621, 627, 649, 655, 661, 667, 673, 687, 693, 699, 705, 711

7. Look for and make use of structure. In many lessons. Some examples are: 49, 55, 119, 131, 143, 155, 193, 255, 273, 279, 285, 291, 297, 361, 367, 373, 379, 385, 397, 511, 517, 523, 529, 535, 541, 547, 553, 559, 573, 579, 603

8. Look for and express regularity in repeated reasoning. In many lessons. Some examples are: 131, 143, 155, 205, 211, 217, 255, 347, 367, 379, 403, 415, 453, 459, 465, 471, 511, 523, 535, 547, 553, 559, 609, 705, 711

Math Story, 1–8, 481–488, 637–644

Measurement
heights, compare, 655–658, 637–644
lengths, compare, 649–652, 673–676
weights, compare, 667–670, 673–676

Mid-Chapter Checkpoint, 34, 96, 140, 202, 252, 332, 394, 450, 526, 600, 664, 702

Minus, 311–314, 317–320, 323–326, 329–332, 335–338, 341–344, 347–350

Modeling
addition, 249–251, 273–276, 279–282, 285–288, 291–294, 297–300
numbers. *See* Numbers
put together, 237–240
shapes, 609–612
subtraction, 329–331
take apart, 317–320, 329–331
take from, 311–314, 323–326, 336–338, 341–344

Next to, 621–624

Nine
compare to, 167–170
count, 155–158, 161–164
model, 155–158
write, 161–164

Nineteen
compare, 409–412, 415–418
count, 409–412, 415–418
model, 409–412
write, 415–418

Numbers
compare. *See* Compare
eight, 143–146, 149–152, 285–288
eighteen, 409–412, 415–418
eleven, 361–364, 367–370
fifteen, 385–388, 391–394
fifty, 453–456
five, 37–40, 43–46, 273–276
four, 25–28, 31–33
fourteen, 373–376, 379–382
to nine, 167–170
nine, 155–158, 161–164, 291–294
nineteen, 409–412, 415–418
one, 13–16, 19–22
to one hundred, 459–462, 465–468
order. *See* Order
pairs of, 273–276, 279–282, 285–288, 291–294, 297–300
same, 55–58, 81–84
seven, 131–134, 137–139, 279–282
seventeen, 397–400, 403–406
six, 119–122, 125–128, 279–282
sixteen, 397–400, 403–406

© Houghton Mifflin Harcourt Publishing Company

ten, 181–184, 187–190, 199–201,
205–208, 211–214, 297–300
three, 25–28, 31–33
twelve, 361–364, 367–370
twenty, 429–432, 435–438, 441–444
two, 13–16, 19–22
use to 15, 391–393
ways to make 5, 49–52
ways to make 10, 193–196
write, 19–22, 31–33, 43–46, 67–70,
125–128, 137–139, 149–152,
161–164, 187–190, 367–370,
379–382, 385–388, 403–406,
415–418, 435–438
zero, 61–64, 67–70

One/Ones
count, 13–16, 19–22, 361–364
draw, 13–16, 22
model, 13–16, 361–364
write, 19–22

On Your Own. *See* Problem Solving

Order
numbers
to five, 55–58
to ten, 199–202
to twenty, 441–444

P

Pair, 49–52

Personal Math Trainer. In all Student
Edition lessons. Some examples are:
10, 75, 76, 78, 113, 114, 116, 175, 176,
178, 225, 226, 228, 304, 305, 308, 353,
356, 358, 422, 423, 426, 479, 480, 490,
566, 568, 570, 633, 634, 646, 680, 681,
684, 718, 719

Picture Glossary. *See* Glossary

Plus, 237–240, 243–246, 249–251,
255–258, 261–264, 267–270, 273–276,
279–282, 285–288, 297–300

Position words
above, 615–618
behind, 627–630
below, 615–618
beside, 621–624
in front of, 627–630
next to, 621–624

Practice and Homework. In every
Student Edition lesson. Some
examples are: 185–186, 247–248,
321–322, 445–446, 515–516, 601–602,
671–672

Problem Solving
activity, 22, 46, 52, 58, 184, 190, 196,
270, 344, 350, 462, 468, 496, 502,
508, 514, 520, 532, 538, 556, 576,
690, 696, 708
On Your Own, 64, 170, 208, 562, 606
Real World activity, 16, 28, 40, 70, 84,
90, 108, 122, 128, 134, 146, 152,
158, 164, 214, 220, 234, 240, 258,
264, 276, 282, 288, 294, 300, 314,
320, 338, 364, 370, 376, 382, 388,
400, 406, 412, 418, 432, 438, 444,
456, 474, 544, 550, 582, 588, 594,
612, 618, 624, 630, 652, 658, 670,
676, 714
Real World On Your Own, 102, 246,
326
Real World Unlock the Problem, 61,
99, 167, 243, 323, 391, 603, 661
strategies
act out addition problems, 243–245
act out subtraction problems,
323–325
draw a picture, 167–170, 391–393,
559–562, 661–663
make a model, 61–64, 99–102,
205–208, 447–449
Unlock the Problem, 205, 447, 559

Put Together, 237–240, 249–251, 255–258

Reading
numbers. *See* Numbers

© Houghton Mifflin Harcourt Publishing Company

© Houghton Mifflin Harcourt Publishing Company

© Houghton Mifflin Harcourt Publishing Company

© Houghton Mifflin Harcourt Publishing Company